Exploring Telescopes:

A Handbook to Observing the Universe from Orbit

Acknowledgment

No journey is ever walked alone, and this book is no exception.

To those who stood beside me—whether in encouragement, challenge, or quiet support—you have shaped this work more than you know.

To the voices that inspired me, the books that sparked new ideas, and the conversations that led me down unexpected paths—thank you for being part of this creative process.

To my family, whose love remains my anchor.
To my friends, who never let me settle for less than my best.

To the readers—without you, this book is just ink on paper. It's your engagement that gives it meaning.

This book exists because of all of you.

Preface

Some books are planned. Others unfold unexpectedly, revealing themselves along the way. This book belongs to the latter category.

It began as a series of thoughts, questions, and discoveries—pieces that, over time, took shape into something larger. My goal is not to provide rigid Insightful Reflections, but rather to offer ideas that provoke thought, encourage exploration, and invite reflection.

Whether you came here seeking knowledge, inspiration, or something in between, I hope this book meets you where you are.

Now, let's begin.

Dedication

To those who walk unseen but leave a lasting impact.

To the dreamers who refuse to be silenced.

To the ones who find courage in the face of doubt.

This book is for you.

Copyright

© All rights reserved.

No part of this book may be copied, distributed, or reproduced in any form without prior written permission from the author or publisher, except for brief excerpts used in reviews or academic references.

Creativity thrives when it is respected—please support original work.

TABLE OF CONTENT

Acknowledgment 2

Preface 3

Dedication 4

Copyright 5

TABLE OF CONTENT 6

CHAPTER 1: THE EVOLUTION OF SPACE TELESCOPES 8

CHAPTER 2: HOW SPACE TELESCOPES WORK .. 15

CHAPTER 3: THE HISTORY OF SPACE TELESCOPES 24

CHAPTER 4: HOW SPACE TELESCOPES WORK .. 33

CHAPTER 5: THE POWER OF SPACE TELESCOPE IMAGES 42

CHAPTER 6: TELESCOPE DESIGN AND ENGINEERING: MAKING THE IMPOSSIBLE POSSIBLE 50

CHAPTER 7: OPERATING AND MAINTAINING SPACE TELESCOPES: CHALLENGES AND INNOVATIONS 62

Chapter 8: The Role of Space Telescopes in Cosmology 74

Chapter 9: The Challenges of Operating Space Telescopes 83

Chapter 10: Future Space Telescopes: The Next Generation of Observatories 92

Chapter 11: Space Telescopes and Citizen Science 102

Chapter 12: The Impact of Space Telescopes on Humanity 112

CHAPTER 1: THE EVOLUTION OF SPACE TELESCOPES

The quest to see beyond Earth's atmosphere is as old as astronomy itself. For centuries, ground-based telescopes served as humanity's window to the cosmos, but they were limited by atmospheric distortion, weather conditions, and light pollution. The idea of placing telescopes in space—beyond these earthly obstacles—was a vision that took decades to realize. From the earliest conceptual musings to the launch of the Hubble Space Telescope and beyond, the evolution of space telescopes is a story of scientific ambition, engineering breakthroughs, and a relentless desire to see the universe with unprecedented clarity.

The problem with ground-based telescopes is the very air we breathe. The Earth's atmosphere, though essential for life, is a nightmare for astronomers. Light from distant celestial objects gets scattered and distorted as it passes through the layers of air, causing the familiar twinkling of stars. This atmospheric turbulence blurs images, making it difficult to capture fine details. Even worse, certain wavelengths of light—like ultraviolet, X-rays, and much of the infrared spectrum—are absorbed entirely by the atmosphere, preventing astronomers from seeing the universe in these critical bands. It's as if nature put a filter on the cosmos, only allowing a fraction of the information to reach us.

The first person to seriously propose putting a telescope in space was Lyman Spitzer, an American astrophysicist. In 1946, Spitzer wrote a paper outlining the advantages of an extraterrestrial observatory. He argued that a telescope placed above the atmosphere would capture images of unprecedented sharpness and could observe celestial objects in wavelengths that are otherwise blocked. But in the 1940s, launching something as complex as a telescope into space was science fiction. The technology simply didn't exist. Rockets were still in their infancy, and the world was just beginning to explore the possibilities of space travel.

The dawn of the space age in the 1950s and 60s changed everything. The launch of Sputnik in 1957 proved that artificial satellites could orbit the Earth. Shortly after, space agencies began considering the possibility of placing telescopes in space. The first significant step came with NASA's **Orbiting Astronomical Observatory (OAO)** program. In 1966, **OAO-1** was launched, but it failed due to power system malfunctions. However, the second attempt, **OAO-2**, launched in 1968, was a success. It carried ultraviolet telescopes, allowing astronomers to study stars and galaxies in a way never before possible. It was a groundbreaking moment—proof that space telescopes could work.

Throughout the 1970s, the idea of a large, versatile space telescope gained momentum. NASA, in collaboration with the European Space Agency (ESA), developed plans for what would become the **Hubble Space Telescope**. This was an ambitious project, aiming to place a telescope with a 2.4-meter primary mirror into low Earth orbit. However, getting Hubble off the ground was no easy task. The project faced delays, budget overruns, and engineering challenges. When it was finally launched aboard the Space Shuttle Discovery in 1990, scientists eagerly awaited crystal-clear images of the cosmos. Instead, they were met with a disaster. Hubble's primary mirror had been ground incorrectly, leading to blurred images. The embarrassment was immense—after decades of planning and billions of dollars spent, the telescope was flawed.

But Hubble's story didn't end there. In 1993, NASA sent astronauts to fix the telescope in one of the most daring space repair missions ever attempted. Using the Space Shuttle Endeavour, astronauts installed corrective optics, essentially giving Hubble a new pair of glasses. When the first corrected images came back, they were breathtaking. Galaxies billions of light-years away, the birthplaces of stars, the intricate details of nebulae—Hubble had finally fulfilled its promise.

The success of Hubble paved the way for a new era of space telescopes. In the years that followed, specialized telescopes were launched to observe the universe in different wavelengths. The **Compton Gamma Ray Observatory**, launched in 1991, studied high-energy cosmic phenomena like black holes and gamma-ray bursts. The **Chandra X-ray Observatory**, launched in 1999, revealed the violent and energetic processes in the universe, from supernova remnants to hot gas clouds surrounding galaxies. The **Spitzer Space Telescope**, launched in 2003, observed the cosmos in infrared light, uncovering hidden star-forming regions and distant exoplanets.

The idea of a next-generation space telescope to surpass Hubble was already in motion long before Hubble's repairs. By the early 2000s, NASA and international partners began developing what would become the **James Webb Space Telescope (JWST)**. Unlike Hubble, which observes primarily in visible and ultraviolet light, JWST was designed to be an infrared telescope, capable of peering into the earliest epochs of the universe. Its massive, segmented mirror—six times the collecting area of Hubble's—was engineered to fold and unfold in space, an unprecedented feat of engineering.

JWST faced its own share of challenges. Originally slated for launch in 2007, it was repeatedly delayed due to budget concerns, technical hurdles, and the complexity of its design. Engineers had to ensure that every component of the telescope would function perfectly in the unforgiving environment of space, one million miles away from Earth—far beyond the reach of repair missions. When JWST finally launched on December 25, 2021, the world held its breath. Its deployment sequence was nerve-wracking, involving over 300 single-point failures that could have doomed the mission. But one by one, the components unfolded perfectly. When the first images were released in July 2022, they were nothing short of revolutionary. Galaxies from the early universe, star-forming regions in exquisite detail, and even the atmospheres of exoplanets—JWST had exceeded expectations.

Today, space telescopes continue to evolve. Future missions aim to build on the successes of Hubble and JWST. The **Nancy Grace Roman Space Telescope**, slated for launch in the late 2020s, will survey the sky with a wide field of view, studying dark energy and detecting exoplanets on an unprecedented scale. Concepts for even larger telescopes, like the **Large Ultraviolet Optical Infrared Surveyor (LUVOIR)** and the **Habitable Worlds Observatory**, envision observatories that could directly image Earth-like exoplanets, searching for signs of life beyond our solar system.

The journey from early ground-based observations to modern space telescopes is a testament to human ingenuity and perseverance. Each new mission builds on the successes and lessons of its predecessors, pushing the boundaries of what is possible. The evolution of space telescopes is not just a story of technology—it's a story of our desire to understand the universe, to see deeper into space and further back in time.

With every new telescope launched, we unlock more secrets of the cosmos, revealing a universe more complex, beautiful, and mysterious than we ever imagined.

CHAPTER 2: HOW SPACE TELESCOPES WORK

Observing the universe from space is no small feat. Unlike ground-based telescopes, which sit comfortably on observatory mounts, space telescopes must function in the harsh vacuum of space, orbiting Earth or stationed at distant Lagrange points. They must endure extreme temperatures, cosmic radiation, and micrometeoroid impacts—all while maintaining precise alignment and transmitting data across vast distances. These engineering marvels rely on advanced optics, cutting-edge detectors, and sophisticated communication systems to capture and relay images of the cosmos. Understanding how space telescopes work requires a deep dive into their structure, operational mechanisms, and the scientific instruments that allow them to reveal the universe in unprecedented detail.

At the heart of every space telescope is its **optical system**, responsible for collecting and focusing light. Most space telescopes operate using a **reflecting telescope** design, where a large primary mirror gathers light and directs it toward a secondary mirror, which then focuses it onto scientific instruments. The size of the primary mirror is crucial—it determines how much light the telescope can collect and, therefore, how faint and distant the observed objects can be. A larger mirror results in sharper images and a greater ability to detect dim celestial bodies. The **Hubble Space Telescope** (HST), for example, has a 2.4-meter mirror, whereas the **James Webb Space Telescope** (JWST) boasts a massive 6.5-meter segmented mirror, allowing it to capture faint infrared signals from the early universe.

Unlike traditional ground-based telescopes, space telescopes must also compensate for the unique challenges of the space environment. Without an atmosphere to stabilize temperatures, space telescopes experience extreme thermal variations. Sun-facing surfaces can heat up to hundreds of degrees, while shadowed areas drop to near absolute zero. To mitigate these effects, telescopes use specialized thermal insulation, radiators, and **sunshields**. JWST's five-layer sunshield, for instance, keeps its instruments at frigid temperatures, ensuring that infrared observations remain unaffected by the heat of the Sun.

Space telescopes operate across a range of **electromagnetic wavelengths**, each revealing different aspects of the cosmos. Some telescopes, like Hubble, are designed for **visible and ultraviolet light**, providing detailed images similar to what the human eye would perceive but with far greater resolution. Others, like JWST, specialize in **infrared light**, allowing them to penetrate cosmic dust and observe the earliest galaxies. The **Chandra X-ray Observatory** focuses on high-energy **X-rays**, detecting the violent processes surrounding black holes and supernova remnants. The choice of wavelength determines the telescope's scientific focus and dictates the type of instruments it requires.

To convert incoming light into usable scientific data, space telescopes employ highly sensitive **detectors**. Unlike traditional photographic plates, modern telescopes use **charge-coupled devices (CCDs)** for visible and ultraviolet imaging, **bolometers** for infrared detection, and **scintillators** for gamma-ray observations. These detectors record photons with extreme precision, translating cosmic signals into digital data that can be analyzed by astronomers. In many cases, telescopes use multiple detectors to cover a broad range of wavelengths simultaneously, maximizing their scientific output.

One of the greatest challenges of operating a space telescope is **stability and precision**. Unlike ground-based observatories, which can be manually adjusted, space telescopes must autonomously maintain perfect alignment. Any slight misalignment—caused by gravitational forces, mechanical vibrations, or even micrometeoroid impacts—can blur images and reduce the quality of data. To counteract this, space telescopes use **reaction wheels**, **gyroscopes**, and **star trackers** to maintain a stable orientation. Hubble, for instance, relies on an array of gyroscopes that allow it to point with extraordinary precision, keeping it locked onto distant galaxies for hours or even days to collect long-exposure images.

Data transmission is another critical component. Space telescopes are often positioned thousands or even millions of kilometers from Earth, requiring powerful **radio antennas** to send and receive signals. Most telescopes transmit their observations to ground stations via NASA's **Deep Space Network (DSN)**, a collection of large radio dishes strategically placed around the globe. These ground stations receive the data, which is then processed and distributed to astronomers worldwide. The amount of data collected can be staggering—JWST, for example, sends back over 50GB of scientific data every day, requiring advanced compression and storage techniques.

Because space telescopes operate in an unforgiving environment, they must be designed for longevity and resilience. Many telescopes, like Hubble, were built with **modular components**, allowing for maintenance and repairs by astronauts. Over its lifetime, Hubble has received multiple servicing missions that replaced its gyroscopes, cameras, and spectrographs, effectively extending its operational lifespan by decades. However, telescopes positioned beyond low Earth orbit, such as

JWST or Chandra, do not have this luxury. Once deployed, they must function autonomously for their entire mission duration, requiring redundant systems and fail-safes to ensure reliability.

One of the most crucial scientific instruments onboard space telescopes is the **spectrograph**. While images provide stunning visual representations of celestial objects, spectrographs break down light into its component wavelengths, revealing essential information about the composition, temperature, and motion of astronomical bodies. By analyzing spectral lines, astronomers can determine what elements make up distant stars, measure the redshift of galaxies, and even detect the atmospheres of exoplanets. Spectroscopy has been fundamental in understanding the nature of dark matter, dark energy, and the large-scale structure of the universe.

Some space telescopes are designed to work in concert with ground-based observatories, creating a **multi-wavelength approach** to astronomy. For example, an event detected in visible light by Hubble might also be studied in X-rays by Chandra and in radio waves by the **Atacama Large Millimeter Array (ALMA)** on Earth. This complementary approach allows scientists to build a more comprehensive understanding of cosmic events, from supernova explosions to black hole accretion disks.

Despite their advanced design, space telescopes are not immune to challenges. Mechanical failures, cosmic ray damage, and software glitches can jeopardize missions. In some cases, telescopes enter **safe mode**, a state in which operations are paused until engineers diagnose and resolve the issue. For instance, Hubble has entered safe mode multiple times due to gyroscope failures, requiring remote troubleshooting from NASA teams. Given the high cost and complexity of these missions, space agencies meticulously plan for contingencies, ensuring that telescopes can continue operating even in the face of unexpected setbacks.

Future space telescopes are pushing the boundaries of what is possible. Concepts like the **LUVOIR** and **Habitable Worlds Observatory** envision telescopes with even larger mirrors and advanced coronagraphs capable of directly imaging Earth-like exoplanets. These next-generation telescopes will use **adaptive optics** and **active wavefront control** to correct for any optical distortions, achieving levels of clarity that surpass even JWST. Some proposals even suggest **modular space telescopes**, where robotic systems could assemble observatories in orbit, allowing for larger and more ambitious designs than what can currently be launched as a single unit.

Understanding how space telescopes work is essential to appreciating their scientific contributions. Every component—from mirrors and detectors to stabilization systems and communication networks—plays a vital role in unlocking the mysteries of the universe. These extraordinary instruments are not just cameras in space; they are time machines, peering into the distant past, revealing the birth and evolution of galaxies, and searching for signs of life beyond Earth. The more we refine these technologies, the clearer our view of the cosmos becomes, bringing us ever closer to answering the deepest questions about our place in the universe.

CHAPTER 3: THE HISTORY OF SPACE TELESCOPES

The quest to see beyond the limits of our atmosphere has been a driving force in astronomy for centuries. While ground-based telescopes have long allowed astronomers to study the stars, the limitations imposed by Earth's atmosphere—blurring effects, light pollution, and absorption of key wavelengths—have always been an obstacle. The dream of placing telescopes in space, free from these restrictions, began as early as the 20th century, long before the technology existed to make it a reality. The development of space telescopes represents a fusion of human ingenuity, engineering brilliance, and the relentless pursuit of knowledge, leading to some of the most groundbreaking discoveries in modern astronomy.

The idea of space-based astronomy was first seriously proposed in the early 1900s. The renowned German rocket scientist Hermann Oberth, one of the fathers of rocketry, speculated about using telescopes in space to escape atmospheric distortions. In his 1923 book *Die*

Rakete zu den Planetenräumen (*The Rocket into Planetary Space*), Oberth described a theoretical observatory in orbit, a vision that would take nearly a century to fully materialize. However, at the time, rocketry was in its infancy, and the very notion of launching telescopes beyond Earth's atmosphere remained a distant dream.

The mid-20th century saw significant advances in both rocketry and observational astronomy, paving the way for the first space-based telescopic instruments. The launch of the Soviet **Sputnik 1** in 1957 marked the beginning of the Space Age, demonstrating that artificial satellites could be placed in orbit. This breakthrough spurred the United States and other nations to consider scientific applications of spaceflight, including astronomical observations free from the distortions of the atmosphere.

In the 1960s, NASA took its first steps toward space-based astronomy with the **Orbiting Astronomical Observatory (OAO)** program. The OAO series was a set of satellites equipped with ultraviolet and X-ray telescopes, designed to observe celestial phenomena that were invisible from Earth's surface. The first OAO, launched in 1966, failed shortly after reaching orbit due to technical malfunctions. However, its successor, **OAO-2**, successfully launched in 1968 and provided unprecedented ultraviolet observations of stars and galaxies. This mission demonstrated the immense potential of space telescopes, proving that placing telescopes above Earth's atmosphere could revolutionize our understanding of the cosmos.

The success of the OAO program laid the groundwork for what would become the most famous space telescope in history: the **Hubble Space Telescope (HST)**. The idea for Hubble originated as early as the 1940s, with astronomers such as Lyman Spitzer advocating for an optical telescope in orbit. However, it was not until the 1970s that serious planning began. NASA, in collaboration with the European Space Agency (ESA), developed an ambitious project to launch a large, long-term space observatory. The telescope, named after Edwin Hubble, the astronomer who discovered the expansion of the universe, was designed to be serviceable by astronauts, allowing for in-orbit maintenance and upgrades.

Hubble was originally scheduled for launch in 1983, but delays and budget constraints pushed it back until April 24, 1990, when it was finally carried into orbit aboard the **Space Shuttle Discovery**. However, shortly after deployment, a devastating flaw was discovered: Hubble's primary mirror suffered from **spherical aberration**, a tiny but critical manufacturing error that caused blurred images. This was a major embarrassment for NASA, as Hubble had been touted as the most advanced telescope ever built.

Fortunately, because Hubble had been designed for servicing, NASA was able to mount a repair mission. In 1993, the **Space Shuttle Endeavour** carried astronauts to Hubble, where they installed corrective optics, effectively giving the telescope "glasses" to correct the flaw. With its vision restored, Hubble began producing some of the most breathtaking and scientifically valuable images ever captured, from the **Hubble Deep Field**, which revealed thousands of distant galaxies, to detailed studies of exoplanets, nebulae, and black holes.

While Hubble revolutionized optical astronomy, it was not the only space telescope to expand our view of the universe. During the 1990s and early 2000s, NASA launched a series of specialized telescopes, each designed to observe different wavelengths of light. These included the **Chandra X-ray Observatory**, launched in 1999 to study high-energy phenomena such as black holes and supernovae, and the **Spitzer Space Telescope**, an infrared observatory that revealed the birth of stars hidden behind cosmic dust. Together, these telescopes formed NASA's **Great Observatories program**, a coordinated effort to study the universe across the full electromagnetic spectrum.

As technology improved, the next generation of space telescopes was developed, culminating in the launch of the **James Webb Space Telescope (JWST)** in 2021. Originally proposed in the late 1990s as a successor to Hubble, JWST faced numerous delays and budget overruns, leading some to doubt whether it would ever fly. However, the telescope was successfully launched on December 25, 2021, aboard an **Ariane 5** rocket. Unlike Hubble, which orbits Earth, JWST was sent to **Lagrange Point 2 (L2)**, a stable gravitational point nearly **1.5 million kilometers from Earth**, where it could operate free from interference.

JWST's capabilities far surpass those of Hubble. Its massive **6.5-meter primary mirror**, constructed from lightweight **beryllium segments**, allows it to collect much more light, making it ideal for studying the faintest and most distant objects in the universe. Additionally, JWST is optimized for **infrared astronomy**, enabling it to peer through dust clouds and observe the formation of the first galaxies after the Big Bang. Within its first year of operation, JWST delivered groundbreaking results, including **detailed spectra of exoplanet atmospheres**, **high-resolution images of star-forming regions**, and **new insights into the early universe**.

Beyond Hubble and JWST, numerous other space telescopes have played crucial roles in expanding our cosmic knowledge. The **Kepler Space Telescope**, launched in 2009, revolutionized our understanding of exoplanets by detecting thousands of planets orbiting distant stars. Before Kepler, only a handful of exoplanets had been confirmed, but its discoveries reshaped our view of the galaxy, suggesting that **Earth-like planets may be common**.

The **Gaia mission**, launched by the **European Space Agency (ESA)** in 2013, has mapped the precise positions and motions of over **a billion stars**, creating the most detailed star catalog in history. This data has provided unparalleled insights into the structure and evolution of the Milky Way.

Newer telescopes, like the **Nancy Grace Roman Space Telescope** (scheduled for launch in the late 2020s), aim to build on Hubble's legacy, with a wider field of view and advanced tools for studying dark matter, dark energy, and exoplanets. Meanwhile, concepts for even more ambitious telescopes, such as **LUVOIR** (Large Ultraviolet Optical Infrared Surveyor) and the **Habitable Worlds Observatory**, are being developed with the goal of directly imaging **Earth-like exoplanets** and searching for signs of life.

The history of space telescopes is one of perseverance, technological breakthroughs, and scientific wonder. From the early vision of Hermann Oberth to the incredible discoveries of Hubble, JWST, and beyond, space-based observatories have fundamentally changed our understanding of the cosmos. Each new telescope builds upon the legacy of its predecessors, pushing the boundaries of what is possible and bringing us ever closer to answering some of the most profound questions about our universe: **Where did we come from? Are we alone? What is the ultimate fate of the cosmos?**

The journey is far from over. As technology advances and our ambitions grow, the next generation of space telescopes promises to reveal even deeper cosmic secrets, ensuring that humanity's exploration of the stars will continue for generations to come.

CHAPTER 4: HOW SPACE TELESCOPES WORK

The ability to place telescopes in space has transformed our understanding of the universe, allowing astronomers to peer deeper into cosmic history than ever before. Unlike their ground-based counterparts, space telescopes operate free from Earth's atmospheric interference, enabling them to capture sharper images and detect wavelengths of light that are otherwise absorbed before reaching the surface. While the concept of a telescope floating in space may seem simple at first, these observatories are intricate machines, requiring cutting-edge engineering, advanced optics, and precise positioning to function. The inner workings of space telescopes are a testament to human ingenuity, blending physics, technology, and careful planning to unlock the mysteries of the cosmos.

At their core, all space telescopes share a few fundamental components: a **primary mirror or lens** to collect and focus light, **scientific instruments** to analyze incoming data, and a **system for maintaining orientation and communication with Earth**. However, beyond these basic elements, the design of each space telescope is highly specialized depending on its mission. Some, like the **Hubble Space Telescope (HST)**, operate primarily in visible and ultraviolet light, while others, such as the **James Webb Space Telescope (JWST)**, are optimized for infrared observations. The specific configuration of each telescope is determined by its scientific goals, the challenges of operating in space, and the limits of current technology.

The most critical component of any telescope is its **optical system**. This system consists of a primary mirror (or lens, in the case of refracting telescopes) that gathers and focuses light. In ground-based telescopes, larger mirrors provide better resolution, but in space, mirror size is constrained by launch vehicle capacity. **Hubble**, for instance, has a **2.4-meter primary mirror**, small compared to the largest ground-based observatories, but capable of producing stunningly detailed images thanks to its position above the atmosphere. Meanwhile, **JWST's**

mirror measures 6.5 meters, giving it significantly greater light-gathering ability. To fit inside its launch rocket, JWST's mirror was designed as a **foldable array of hexagonal segments**, which automatically aligned once the telescope reached its operational orbit.

Optical telescopes typically use one of two designs: **reflecting telescopes**, which use mirrors to focus light, and **refracting telescopes**, which rely on lenses. Most space telescopes use the **reflecting design** because mirrors are lighter and easier to manufacture than large lenses. However, even within reflecting telescopes, different mirror configurations exist. Hubble, for instance, uses a **Cassegrain system**, where light is first collected by a concave primary mirror, then reflected to a secondary mirror that directs it toward scientific instruments. This compact design allows for a long focal length without requiring an extremely large telescope structure.

While optical telescopes operate similarly to those on Earth, telescopes designed to observe **infrared, ultraviolet, X-ray, or gamma-ray radiation** require more specialized instruments. **Infrared telescopes**, such as JWST or the **Spitzer Space Telescope**, must be kept extremely cold to detect faint heat signals from distant objects. If the telescope itself radiates heat, it will interfere with observations. To solve this, infrared telescopes are equipped with cooling systems, often involving **cryogenic components** that use liquid helium or passive cooling shields to maintain ultra-low temperatures. JWST, for instance, employs a **multi-layer sunshield the size of a tennis court** to block heat from the Sun, allowing it to operate at around **-233°C (-387°F)**.

In contrast, **X-ray and gamma-ray telescopes** face a different challenge: these high-energy wavelengths do not behave like visible light, making traditional mirrors ineffective. Instead, X-ray telescopes use specially designed **grazing incidence mirrors**, where incoming photons are gently deflected at shallow angles until they reach a focal point. NASA's **Chandra X-ray Observatory** is one such example, using nested cylindrical mirrors to capture X-rays from black holes and supernovae. Gamma-ray telescopes, like the **Fermi**

Gamma-ray Space Telescope, do not use mirrors at all but instead rely on **particle detectors** that record the interactions of high-energy photons.

Once light is collected, it must be analyzed by a **suite of scientific instruments**. These instruments vary widely depending on the telescope's mission but typically include **cameras**, **spectrographs**, and **imaging systems** optimized for different wavelengths.

1. **Cameras**, such as Hubble's **Wide Field Camera 3**, capture high-resolution images of celestial objects. These cameras often use **CCD (charge-coupled device) sensors**, which convert incoming photons into electronic signals, allowing astronomers to construct digital images of space.
2. **Spectrographs**, like JWST's **Near-Infrared Spectrograph (NIRSpec)**, break light into its component wavelengths, revealing details about an object's composition, temperature, motion, and more. Spectroscopy is a powerful tool that allows

scientists to study exoplanet atmospheres, distant galaxies, and interstellar gas clouds.
3. **Polarimeters** analyze the polarization of light, providing insights into the structure of cosmic magnetic fields and high-energy environments such as black hole jets.

Beyond optics and instruments, space telescopes require **complex control and communication systems** to operate effectively. Unlike ground-based telescopes, which can be adjusted in real-time by astronomers, space telescopes must be **remotely controlled**, relying on automated systems and periodic commands from Earth.

Positioning a space telescope is an intricate task. A telescope floating in space lacks a fixed foundation, meaning that even the slightest force—from solar radiation pressure to micro-meteoroid impacts—can cause it to drift. To counteract this, space telescopes use **reaction wheels and gyroscopes** to make precise adjustments to their orientation. Hubble, for instance, has **four gyroscopes**, which work by exploiting the conservation of angular momentum to maintain its position. If fine-tuning is needed, small **thrusters** or **momentum wheels** can be used to correct the telescope's alignment.

Because space telescopes cannot store unlimited amounts of data, they must regularly **transmit their observations back to Earth**. This is done via **high-frequency radio communications**, often relayed through NASA's **Deep Space Network (DSN)**, a global system of massive ground antennas. These antennas receive the signals, decode the data, and store it for analysis by astronomers worldwide.

Maintaining space telescopes is one of the greatest challenges of their design. Unlike ground-based telescopes, which can be serviced as needed, most space telescopes operate **without human intervention**. This means they must be built to function autonomously for years or even decades. Hubble is one of the rare exceptions—it was designed to be **serviced by astronauts**, leading to multiple repair and upgrade missions between 1993 and 2009. These missions extended Hubble's lifespan far beyond its original expectations, keeping it operational for over 30 years. Newer space telescopes, however, are typically designed to be **fully autonomous**, relying on redundancy systems and fail-safes to extend their longevity.

Future space telescopes will continue to push the limits of what is possible. Concepts such as the **Large Ultraviolet Optical Infrared Surveyor (LUVOIR)** and the **Habitable Worlds Observatory** aim to take space-based astronomy to the next level, with even larger mirrors, better imaging resolution, and new methods for detecting exoplanets. Advances in **adaptive optics, artificial intelligence, and quantum communication** may also play a role in the next generation of space telescopes, allowing for real-time corrections to optical distortions and more efficient data transmission.

The intricate design and operation of space telescopes highlight the immense effort required to explore the universe. Every component, from the primary mirror to the tiny reaction wheels, is meticulously engineered to withstand the harsh environment of space while delivering precise and valuable data. As technology advances and new observatories are launched, space telescopes will continue to revolutionize our understanding of the cosmos, uncovering secrets that remain hidden beyond the limits of the human eye.

CHAPTER 5: THE POWER OF SPACE TELESCOPE IMAGES

Space telescopes have provided us with some of the most breathtaking and scientifically important images of the universe. These images do not simply serve as beautiful works of art, but as invaluable windows into the cosmos, allowing us to witness phenomena that are both stunning and crucial to our understanding of space, time, and the very fabric of reality. The power of space telescope images lies not only in their visual impact but in the rich scientific data they offer. Each image tells a story, whether it's the birth of a star, the collision of galaxies, or the glimpse of light from the earliest moments after the Big Bang.

In the world of astronomy, an image is often more than just a visual representation. It is a complex blend of science, technology, and human curiosity. Space telescopes, with their capacity to observe distant objects in high detail and across various wavelengths, produce images that are the result of intricate processes. These include capturing raw data through multiple exposures, analyzing light at different wavelengths (from ultraviolet to infrared), and using computational techniques to enhance and interpret the images.

One of the most famous and iconic images captured by space telescopes is the **Hubble Deep Field**. This image, taken by the **Hubble Space Telescope (HST)** in 1995, is a long-exposure photograph of a small patch of sky. Incredibly, it revealed thousands of galaxies—some of which are billions of light-years away—shedding light on the vastness of the universe. The sheer number of galaxies and the detail in the image surprised even the most seasoned astronomers, offering profound insights into the structure and history of the universe.

The **Hubble Deep Field** also illustrated a critical concept in astronomy: the idea of the observable universe. This patch of sky, when viewed with a powerful enough telescope, contains an unimaginable number of galaxies, each with its own stars, planets, and potential for harboring life. The image serves as both an awe-inspiring piece of art and a profound scientific tool, allowing astronomers to study galaxy formation and evolution.

However, the Hubble Space Telescope's contributions to our understanding of the universe extend far beyond just the Deep Field. Hubble's observations have been pivotal in answering some of the most profound questions in cosmology, such as the **rate of expansion of the universe** and the **existence of dark energy**. The telescope has observed the **cosmic microwave background radiation**, the faint afterglow of the Big Bang, and studied **exoplanets** in detail, contributing to the search for alien life. Hubble's imagery has also helped to define the age of the universe, provided insights into the life cycle of stars, and even uncovered evidence of the mysterious **dark matter** that makes up a large portion of the universe's mass.

The ability of space telescopes like Hubble to capture images across multiple wavelengths—such as ultraviolet, visible, and near-infrared light—has been revolutionary. Different wavelengths of light provide different types of information about objects in space. For instance, **infrared images** can reveal the cooler, hidden structures of the universe, such as **dust clouds** that block visible light but allow infrared radiation to pass through. **Ultraviolet images**, on the other hand, highlight high-energy phenomena like the formation of new stars or the radiation emitted by active galactic nuclei (AGN).

One of the most exciting space telescope missions in recent years is the **James Webb Space Telescope (JWST)**, which will further extend our ability to capture images across the infrared spectrum. JWST, with its much larger mirror and state-of-the-art instruments, promises to revolutionize our understanding of the universe. It will allow astronomers to look back in time, observing galaxies as they were just a few hundred million years after the Big Bang. The telescope is also expected to provide new insights into the formation of stars and planets and to continue the search for exoplanets, especially those that might be capable of supporting life.

JWST's design, including its **gold-coated mirrors** and its complex cooling system, is specifically optimized for the infrared spectrum. Unlike Hubble, which observes in the visible and ultraviolet wavelengths, JWST's **infrared capability** allows it to see through dust clouds and capture details of objects too faint or distant for other telescopes. This makes it especially powerful for studying distant galaxies, exoplanets, and even the most mysterious objects in the universe, such as the **first black holes** or the **dark energy** responsible for the accelerated expansion of the universe.

Another significant contribution of space telescope images is their role in the study of **star formation**. The **nebulae**—vast clouds of gas and dust—are the birthplaces of stars. The images captured by space telescopes have provided scientists with detailed views of these cosmic nurseries, allowing them to study the processes that lead to the birth of stars and planets. The **Orion Nebula**, for example, is one of the most studied regions in the sky, and the images from Hubble and other space telescopes have allowed scientists to observe the formation of new stars and planetary systems in stunning detail.

In addition to star formation, space telescope images have also shed light on **stellar death**. Supernovae, the explosive deaths of massive stars, have been captured by telescopes like Hubble and the **Chandra X-ray Observatory**. These images have allowed astronomers to study the remnants of these powerful explosions, which can create some of the most exotic and extreme objects in the universe, such as **neutron stars** and **black holes**.

One of the most important lessons from these images is the role of **supernovae** in the creation of heavy elements. These elements, such as gold and uranium, are forged in the heart of dying stars and are scattered across the universe during supernova explosions. Space telescopes have allowed scientists to study these remnants in detail, helping us to understand how the chemical elements essential for life came to be distributed throughout the universe.

The images from space telescopes also play an essential role in the discovery and study of **exoplanets**—planets that orbit stars other than the Sun. The **Kepler Space Telescope**, which was specifically designed to find exoplanets, has provided some of the most detailed and extensive images of distant worlds. Although it did not produce traditional photographs of exoplanets, Kepler's **light curve data**—measuring the dimming of a star as a planet transits in front of it—has led to the discovery of thousands of exoplanets, many of which are located in the **habitable zone** where liquid water might exist.

JWST is also expected to contribute significantly to exoplanet research, particularly by observing the **atmospheres of exoplanets**. By analyzing the way starlight passes through the atmosphere of an exoplanet, scientists can learn about its composition, weather, and even the potential for life. This capability has the potential to revolutionize the search for life beyond Earth, allowing scientists to study planets around distant stars that might host life as we know it or in entirely new forms.

The ability to capture and interpret the images produced by space telescopes is not just about visual beauty or scientific curiosity; it is about using those images to **answer profound questions** about our place in the universe. Every image from a space telescope brings us closer to understanding fundamental aspects of existence: the origins of stars and planets, the history of galaxies, the nature of black holes, and even the potential for life beyond Earth. Each new image invites us to wonder, to question, and to explore.

As we continue to push the boundaries of space exploration, the power of space telescope images will only increase. New telescopes will provide unprecedented views of the universe in wavelengths we have never seen before, opening new windows of discovery. But even as technology advances, one thing will remain constant: the awe and wonder that these images inspire in us all, as we continue our journey to understand the vast and beautiful universe in which we live.

CHAPTER 6: TELESCOPE DESIGN AND ENGINEERING: MAKING THE IMPOSSIBLE POSSIBLE

Designing and building space telescopes is one of the most intricate and challenging tasks in modern science and engineering. A space telescope is not just a simple device that collects light; it is a finely tuned instrument that must overcome a multitude of obstacles in order to observe the universe with unparalleled precision. These obstacles include the extreme conditions of space, the need for precise alignment and calibration, and the requirement for sophisticated technologies to capture and process light from objects billions of light-years away.

When we think about space telescopes like the **Hubble Space Telescope** (HST) or the **James Webb Space Telescope** (JWST), it's easy to imagine them as giant cameras floating in space, snapping pictures of distant galaxies and nebulae. However, the reality of their design is far more complex. These telescopes are the result of decades of research, development, testing, and innovation. The engineers and scientists who create them must think about everything, from the size and shape of the mirrors to how the telescope will communicate with Earth, to how it will survive the hostile environment of space.

To truly understand the magnitude of these achievements, we need to dive into the intricacies of telescope design and engineering. Let's break down some of the most important elements that go into the creation of space telescopes and why they are crucial for successful space observation.

The Heart of the Telescope: The Mirror

The primary mirror of a space telescope is arguably the most critical component in its design. This mirror is responsible for gathering light from distant celestial objects and focusing it onto instruments that will process the data. The size and shape of the mirror have a direct impact on the telescope's ability to observe distant objects with high resolution. The larger the mirror, the more light it can collect, and the sharper the image.

For example, the **Hubble Space Telescope**'s primary mirror is **2.4 meters** in diameter, while the **James Webb Space Telescope** has a much larger mirror, measuring **6.5 meters** across. The size difference between the two mirrors is significant because JWST's larger mirror allows it to collect more light and observe fainter, more distant objects. The increased light-gathering power of a larger mirror is essential for observing the earliest galaxies that formed after the Big Bang, as well as the faint glow of distant stars and exoplanets.

However, building a large mirror is not as simple as just making it bigger. The mirror must be precisely shaped to ensure that it focuses light accurately. A single imperfection, no matter how small, can distort the image and make the telescope useless. This is why the mirror's shape and surface must be polished to an incredible level of precision. For example, the Hubble Space Telescope's mirror was polished to an accuracy of about **1/50th of the thickness of a human hair** before it was launched into space. Even the slightest imperfections could cause serious problems, so the mirror must be flawless.

For the James Webb Space Telescope, the engineering challenge was even greater. JWST's primary mirror is made up of **18 hexagonal segments**, each of which must be carefully aligned once the telescope is deployed in space. This is a huge challenge because the segments must work together as a single unit to form a perfect surface for focusing light. The mirror segments are coated with **a thin layer of gold** to enhance their ability to reflect infrared light. This intricate assembly process requires advanced technologies, including actuators that will adjust each segment's position with micrometer precision once the telescope is in orbit.

The mirror design for space telescopes like Hubble and JWST also takes into account the effects of temperature. The harsh environment of space means that materials undergo extreme thermal expansion and contraction, which can affect the alignment of the mirror. To address this issue, engineers use **materials with low thermal expansion**, such as **beryllium** for the JWST mirror, to minimize the impact of temperature changes.

Instruments for Detection and Analysis

Once light is collected and focused by the mirror, it must be analyzed by the telescope's scientific instruments. These instruments are responsible for measuring different properties of light, such as its wavelength, intensity, and polarization. The choice of instruments depends on the type of observations the telescope is intended to make. Space telescopes can observe light in a range of wavelengths, from **ultraviolet** to **infrared**, and the instruments must be designed to capture data across these spectra.

For example, the Hubble Space Telescope is equipped with a variety of instruments that observe light in the **visible**, **ultraviolet**, and **near-infrared** wavelengths. These include the **Wide Field Camera 3** (WFC3), which can capture high-resolution images across multiple wavelengths, and the **Advanced Camera for Surveys** (ACS), which is designed for detailed surveys of distant galaxies.

The **James Webb Space Telescope**, on the other hand, is optimized for **infrared** observations. Its instruments include the **Near-Infrared Camera (NIRCam)**, which can observe faint infrared light from distant galaxies, and the **Mid-Infrared Instrument (MIRI)**, which is designed to study the cool, distant objects in the universe, such as star-forming regions and exoplanets. JWST's instruments are specifically designed to observe the universe in a way that Hubble cannot, allowing astronomers to look back in time to the earliest moments after the Big Bang.

Space telescopes often require highly sensitive detectors to capture the faint light coming from distant objects. These detectors must be able to pick up low levels of light without being overwhelmed by noise. To achieve this, space telescope detectors often use **charge-coupled devices (CCDs)** or **infrared detectors** that can detect individual photons of light. These detectors are often cooled to very low temperatures to reduce thermal noise and increase their sensitivity.

In addition to the basic scientific instruments, modern space telescopes are equipped with a range of other components that allow them to make precise measurements and navigate the cosmos. For instance, telescopes like Hubble and JWST have **fine guidance systems** that use star trackers to help them stay pointed at a specific region of the sky. These systems ensure that the telescope remains stable and accurately oriented during long exposures, preventing image blur and ensuring the precision required for scientific observations.

Powering the Telescope: Energy in Space

One of the most significant challenges in designing a space telescope is the need to power it in the absence of an electrical grid. Space telescopes rely on **solar panels** to generate electricity from sunlight, which is then used to power the telescope's instruments, communications systems, and other components.

The solar panels are often designed to be highly efficient, using materials like **gallium arsenide** to capture as much sunlight as possible. The amount of power required depends on the size and complexity of the telescope. The **Hubble Space Telescope** uses solar arrays that provide enough power for its instruments and communication systems, but the **James Webb Space Telescope** requires much more power due to its larger size and more advanced instruments. As a result, JWST's solar arrays are significantly larger and more complex, with multiple layers of protective material to shield them from the harsh conditions of space.

In addition to solar power, space telescopes often use **batteries** to store energy for use during times when the telescope is in the shadow of a planet or the Sun. These batteries ensure that the telescope can continue to function even when it is not receiving direct sunlight.

Thermal Control: Surviving Extreme Conditions

Space telescopes must also deal with extreme temperature fluctuations. The **Hubble Space Telescope** orbits the Earth at an altitude of about **547 kilometers** (340 miles), where temperatures can swing between **-150°C** (-238°F) in the shadow of Earth to **+150°C** (302°F) when exposed to direct sunlight. These temperature extremes can cause materials to expand and contract, which could affect the telescope's performance.

To mitigate these temperature fluctuations, space telescopes use a variety of **thermal control systems**. These systems are designed to regulate the temperature of the telescope and ensure that its instruments stay within the optimal range for operation. For instance, Hubble has **multi-layer insulation (MLI)** that covers its exterior and helps to maintain a stable temperature. The **James Webb Space Telescope**, being an infrared telescope, requires an even more sophisticated thermal control system. JWST has a large **sunshield** made of **five layers of special material** that block sunlight and keep the telescope's instruments cool, especially its infrared detectors, which need to be at **very low temperatures** to function properly.

Thermal control systems also ensure that the telescope's mirrors and instruments are kept at the right temperatures. For example, the mirror of the James Webb Space Telescope must be kept at **cryogenic temperatures** to minimize interference from heat sources and ensure that it can capture faint infrared light.

Deployment and Alignment in Space

Once a space telescope is built, it must be carefully launched into space, deployed, and aligned. This process is one of the most complex and delicate stages of the entire mission. Space telescopes are often too large to fit into the rocket's payload bay in their fully extended form, so they must be carefully folded or stowed before launch.

Once the telescope is in orbit, it must be unfolded and aligned. For instance, the **James Webb Space Telescope** was folded up for launch aboard an Ariane 5 rocket and required over a month to fully deploy once it reached its destination at **Lagrange Point 2**. This process involves the deployment of its solar arrays, antennas, sunshield, and mirror segments. Each component must be deployed and positioned with incredible precision, often using motors, actuators, and a series of sensors to monitor the telescope's alignment.

Once deployed, the telescope must also be aligned. This is particularly challenging for telescopes with large mirrors, such as JWST. The mirror segments must be aligned with micrometer precision to ensure that the telescope can focus light accurately. The alignment process for JWST involves adjusting the position of each mirror segment and fine-tuning the telescope's instruments.

Communication and Data Transmission

Once the telescope is operational, it must be able to send data back to Earth. Space telescopes use **high-frequency radio signals** to communicate with ground stations. These signals carry the data collected by the telescope's instruments, as well as status updates and other critical information.

Data transmission can be slow, especially for space telescopes like JWST, which will be located **1.5 million kilometers (1 million miles)** from Earth. This means that data must be carefully compressed and transmitted in small batches. However, the information that is returned from space telescopes is invaluable for advancing our understanding of the universe.

The engineering that goes into designing, building, and operating space telescopes is nothing short of extraordinary. It is a testament to human ingenuity and perseverance, combining the best of science, technology, and engineering to explore the farthest reaches of the cosmos. Through these remarkable achievements, we have been able to peer deeper into the universe than ever before and uncover the secrets of space that were once beyond our reach.

CHAPTER 7: OPERATING AND MAINTAINING SPACE TELESCOPES: CHALLENGES AND INNOVATIONS

Operating and maintaining space telescopes presents some of the most daunting challenges in modern science and technology. Unlike ground-based telescopes, which can be serviced, repaired, or adjusted relatively easily, space telescopes are subjected to the harsh conditions of space, where maintenance is virtually impossible. Once launched and deployed, they must function autonomously for years, with no opportunity for human intervention unless something goes wrong—by which point, any fix is too late. This unique aspect of space telescopes requires that they be designed for long-term reliability and self-sufficiency from the outset.

Space telescopes like the **Hubble Space Telescope** (HST) or the **James Webb Space Telescope** (JWST) provide invaluable insights into the workings of the universe, enabling us to make discoveries that would have been impossible from Earth's surface. However, keeping them operational in the vacuum of space presents a set of challenges that span multiple disciplines, from hardware reliability to data transmission, power management, thermal control, and advanced software systems.

In this chapter, we explore the operational challenges and innovations involved in running these magnificent observatories, from pre-launch testing to day-to-day operations in orbit, and the remarkable strategies employed to ensure their longevity and functionality.

Mission Design and Launch: Ensuring Long-Term Success

The design and launch of a space telescope is the first critical step toward ensuring its successful operation in orbit. The success of a mission depends heavily on meticulous planning, testing, and simulation of the telescope's capabilities and performance. Space telescopes undergo a rigorous and exhaustive process of **pre-launch validation** to test each subsystem and verify that the design can withstand the harsh environment of space. This includes subjecting the telescope to intense vibration tests to simulate the stresses of launch, thermal vacuum testing to mimic the extreme temperature variations in space, and other simulations to ensure that every component performs under realistic conditions.

For example, before its 1990 launch, Hubble underwent extensive ground testing at NASA's Goddard Space Flight Center. Every component was put through harsh conditions—radiation tests, extreme temperature cycling, and high-vibration shaking to simulate the stresses of launch. The objective was to identify any weaknesses in the design and ensure that the telescope could survive the journey through space and function optimally once deployed.

Once the telescope passes these tests, the next challenge is the actual launch. Unlike ground-based telescopes, which can be easily transported and installed, space telescopes must be carefully packaged into rockets that take them to orbit. Space telescopes like Hubble and

JWST are often folded up or stowed to fit into the rocket's payload bay, and the deployment of the telescope in space requires a complex and highly choreographed sequence of events. For example, JWST's deployment, which involved the unfolding of its **sunshield** and the positioning of its mirror segments, took weeks and required high precision, as any failure in the deployment process could result in mission failure.

Operational Challenges in Space: The Space Environment

Once a telescope is successfully launched and deployed, it enters the challenging environment of space, where it must operate in a variety of extreme conditions. Space is a vacuum, meaning there is no air to provide cooling or heat dissipation. Temperatures in space can vary drastically, from near absolute zero when the telescope is in the shadow of a planet or star to extremely high temperatures when it is exposed to direct sunlight. These temperature swings can cause materials to expand or contract, which can affect the performance of the telescope's instruments, mirrors, and delicate sensors.

To manage these extreme temperature fluctuations, space telescopes rely on sophisticated **thermal control systems**. For example, the **James Webb Space Telescope** (JWST) is equipped with a multi-layer **sunshield** that blocks sunlight and shields the telescope from solar radiation, while its mirrors and instruments are cooled to near absolute zero to minimize the interference from thermal noise. Additionally, **heat pipes**, **radiators**, and **thermal insulation** are used to regulate the temperature of sensitive components like detectors and mirrors.

Another critical challenge in space is the lack of gravity. While gravity on Earth allows for easy adjustments to large telescopes, in space, gravity no longer provides a stabilizing force. This means that the telescope must rely on **reaction wheels**, **gyroscopes**, and **star trackers** to maintain its orientation. These systems help the telescope point precisely at specific celestial objects and maintain its stability during long observation sessions. The importance of this pointing accuracy cannot be overstated, as even small errors in alignment can cause blurry images, especially when dealing with distant objects like galaxies or exoplanets.

Space telescopes are also subject to **radiation** from the Sun, cosmic rays, and other sources. This radiation can damage sensitive electronic components and degrade the performance of the telescope over time. To mitigate these risks, space telescopes use **radiation-hardened electronics** and shielding to protect their delicate systems. For example, both Hubble and JWST use radiation shielding around their electronics to reduce the risk of damage, and their instruments are designed to handle the effects of cosmic radiation over long periods.

Monitoring and Control: Commanding the Telescope from Earth

One of the most challenging aspects of operating a space telescope is maintaining communication with the spacecraft from Earth. Space telescopes are located far beyond the reach of astronauts, so all communications are handled remotely using high-frequency **radio signals**. These signals are transmitted between the telescope and ground stations, often located at NASA's **Goddard Space Flight Center** or other space agencies' facilities, which are responsible for receiving data, issuing commands, and controlling the telescope's operations.

This communication system is crucial for two main reasons: controlling the telescope and downloading scientific data. Telemetry, which consists of diagnostic and health information about the telescope, is constantly transmitted back to Earth to ensure that all systems are functioning correctly. This includes information about the telescope's **power levels**, **temperature**, **alignment**, and **pointing accuracy**. If any component malfunctions or shows signs of degradation, the ground team must assess the situation and take corrective action.

Sending commands to the telescope is not as simple as typing instructions into a computer. Communication with space telescopes is often delayed due to the vast distances between the spacecraft and Earth. For example, the **James Webb Space Telescope**, which is located at **Lagrange Point 2 (L2)**, is about **1.5 million kilometers** (1 million miles) away from Earth, meaning it takes about **five seconds** for a signal to travel each way. This delay means that any command sent from Earth takes a minimum of ten seconds to receive a response. This creates a challenge when real-time adjustments or troubleshooting are required. As a result, space telescopes are often equipped with **autonomous systems** that can perform routine functions without needing immediate input from Earth.

Another challenge is the **downloading of scientific data**. Space telescopes are designed to capture vast amounts of data in the form of images, spectra, and other measurements. However, the limited bandwidth of communication channels means that only a fraction of this data can be transmitted back to Earth at any given time. This requires efficient data compression algorithms and a carefully managed schedule for data transmission. For instance, the **Hubble Space Telescope** transmits data back to Earth in scheduled bursts, prioritizing the most important observations. Similar systems are employed by JWST, though with even more complexity due to the larger volume of data being generated.

Maintaining Space Telescopes: Innovations in Autonomy and Self-Sufficiency

Since space telescopes cannot be easily repaired or maintained by astronauts, engineers have had to develop innovative strategies for ensuring their longevity and functionality. These strategies focus on **self-sufficiency** and **autonomy**.

One of the key innovations in space telescope design is the development of systems that can **detect and correct problems autonomously**. For example, the **Hubble Space Telescope** had to contend with a flaw in its mirror shortly after its launch. The primary mirror of the telescope was polished to the wrong shape, resulting in blurry images. Fortunately, engineers were able to develop corrective optics that were installed during a servicing mission in 1993. While this mission is often cited as an example of human intervention in space, it was also the turning point in the design of space telescopes, making the industry realize the importance of designing systems that could fix themselves in the absence of astronauts.

The **James Webb Space Telescope**, which launched in December 2021, is built with a high degree of **autonomous control**. One example is its ability to adjust its mirror segments using **micro-adjustment motors**. These motors, which are capable of making minute adjustments to the mirror's position, can perform the fine alignment required for perfect image quality without needing to be manually adjusted. In addition, JWST is equipped with a **self-correcting navigation system** that uses **star trackers** and **gyroscopes** to autonomously maintain its orientation in space. The telescope can automatically make adjustments to its orientation based on real-time data, reducing the risk of malfunctions.

In addition to autonomy, space telescopes are equipped with **redundant systems**. These backup systems are designed to take over in case of a failure in the primary system. For example, space telescopes often have **backup power systems**, **redundant gyroscopes**, and **duplicate sensors** to ensure that the telescope can continue to operate even if one component fails. Redundancy is essential for maintaining the telescope's functionality over the course of its mission, as many space telescopes are designed to operate for **decades**.

End of Life and Final Operations

Eventually, all space telescopes will reach the end of their operational life. When this happens, they are either decommissioned or repurposed. The **Hubble Space Telescope**, for example, is expected to continue operating for several more years, but when it finally reaches the end of its life, it will be retired, and its mission will end. Hubble's decommissioning process includes plans for it to remain in orbit for some time before re-entering the Earth's atmosphere, where it will burn up. Other space telescopes, such as JWST, are designed to remain in their designated orbits for decades.

In Insightful Reflection, the operation and maintenance of space telescopes require exceptional levels of innovation, autonomy, and foresight. These spacecraft are designed to operate without human intervention, using advanced systems and cutting-edge technology to ensure that they continue to function throughout their missions. As the demand for scientific exploration grows and new telescopes are launched, the challenges of operating these instruments will only continue to evolve. Nevertheless, the work that has been done in developing the tools, systems, and strategies for maintaining space telescopes has already provided the world with countless breakthroughs in our understanding of the cosmos.

Chapter 8: The Role of Space Telescopes in Cosmology

The role of space telescopes in cosmology is both profound and transformative. These instruments, which are positioned beyond Earth's atmosphere, have opened up the universe in ways that ground-based telescopes never could. From observing the most distant galaxies to unraveling the mysteries of dark energy and dark matter, space telescopes are essential tools in the quest to understand the cosmos. They have redefined what we know about the universe's origins, structure, and fate. This chapter will explore the critical role that space telescopes play in cosmology, highlighting their contributions to our understanding of the universe and their significance in shaping future discoveries.

One of the primary reasons space telescopes are so vital to cosmology is their ability to observe the universe without the interference of Earth's atmosphere. Ground-based telescopes, though immensely powerful, are often limited by atmospheric distortion, which can blur the light from distant celestial objects. In contrast, space telescopes can capture clear, sharp images across various wavelengths of light, from visible to infrared and ultraviolet, enabling scientists to study cosmic phenomena in unprecedented detail.

In the early days of astronomy, astronomers had to rely on simple optical telescopes, and their view of the universe was limited to the visible spectrum of light. However, the advent of space-based observatories in the latter half of the 20th century opened new windows to the cosmos. With the launch of the Hubble Space Telescope in 1990, the study of cosmology was forever changed. Hubble not only provided breathtaking images of distant galaxies but also made groundbreaking contributions to the understanding of the universe's expansion, helping to measure the rate of this expansion with remarkable precision.

The concept of an expanding universe, first proposed by Edwin Hubble in the 1920s, was largely based on observations made with ground-based telescopes.

However, Hubble's measurements were initially somewhat limited by the technological constraints of the time. It wasn't until the launch of the Hubble Space Telescope that scientists were able to refine these measurements, offering more accurate estimates of the Hubble constant—the value that describes the rate of expansion of the universe. By measuring the redshift of galaxies and comparing their distances, Hubble provided a clearer picture of how fast the universe is expanding.

Space telescopes have also played a crucial role in the study of cosmic background radiation. One of the most significant discoveries in cosmology came in 1965 when Arno Penzias and Robert Wilson accidentally detected the cosmic microwave background (CMB) radiation. This faint afterglow of the Big Bang is a relic from the early universe, and its study has provided crucial insights into the universe's birth and evolution. While ground-based telescopes can detect some aspects of this radiation, it was not until the launch of the Cosmic Background Explorer (COBE) satellite in 1989, followed by the Wilkinson Microwave Anisotropy Probe (WMAP) in 2001, that scientists were able to map the CMB in exquisite detail. These space missions have been instrumental in confirming the Big Bang theory of the universe's origin, helping cosmologists understand the first moments of the universe and the processes that led to its current structure.

A critical aspect of modern cosmology involves understanding the composition of the universe—what it's made of and how its components interact. It is now widely accepted that ordinary matter, the stuff we see around us, makes up only about 5% of the total mass and energy in the universe. The remaining 95% consists of dark matter and dark energy, two mysterious and largely invisible substances that have yet to be directly detected. However, space telescopes have provided indirect evidence for both dark matter and dark energy, and they are essential tools for studying these enigmatic components.

Dark matter, for example, exerts a gravitational influence on visible matter, and its presence can be inferred by observing the motion of galaxies and galaxy clusters. The study of galaxy rotation curves—how stars move within galaxies—has shown that galaxies rotate much faster than expected based on the visible matter alone. This discrepancy can only be explained if there is an unseen mass in galaxies, which we call dark matter. The Hubble Space Telescope has played a critical role in studying galaxy clusters and gravitational lensing—an effect where light from a distant object is bent by the gravitational field of a massive object in the foreground. This effect allows astronomers to map the distribution of dark matter in galaxy clusters, providing important clues about its nature.

On the other hand, dark energy is even more mysterious. This force, which is believed to be responsible for the accelerated expansion of the universe, remains one of the greatest challenges in cosmology. The discovery of dark energy in 1998, made through observations of distant supernovae, was one of the most unexpected and groundbreaking revelations in modern cosmology. Space telescopes such as the Hubble Space Telescope have been essential in studying the effects of dark energy, as they allow astronomers to observe distant supernovae and galaxies and measure their redshifts with great accuracy. By studying the distribution of galaxies and the rate at which the universe's expansion is accelerating, cosmologists are hoping to uncover more about the nature of dark energy and its role in the fate of the universe.

The role of space telescopes in cosmology extends beyond the observation of distant galaxies and cosmic phenomena. They also play a key role in studying the very fabric of the universe itself. One of the major questions in cosmology is the nature of space-time—how it behaves, stretches, and curves in the presence of mass and energy. The theory of general relativity, developed by Albert Einstein in the early 20th century, predicts that massive objects like black holes and neutron stars should warp the fabric of space-time. Space telescopes are crucial for studying these extreme environments, where the laws of physics are pushed to their limits.

One of the most significant breakthroughs in this area came with the detection of gravitational waves in 2015, a phenomenon predicted by general relativity. Gravitational waves are ripples in space-time caused by the acceleration of massive objects, such as the collision of black holes. While ground-based detectors like LIGO (Laser Interferometer Gravitational-Wave Observatory) are designed to detect these waves, space telescopes like the planned Laser Interferometer Space Antenna (LISA) will be able to detect lower-frequency gravitational waves from more massive cosmic events. This will provide a unique view of the universe, allowing scientists to study phenomena such as black hole mergers, neutron star collisions, and even the earliest moments of the universe's existence.

Space telescopes are also key in the search for exoplanets—planets that orbit stars beyond our solar system. The discovery of exoplanets has been one of the most exciting developments in modern astronomy, and space-based telescopes have played a central role in this field. Instruments like the Kepler Space Telescope have been able to detect thousands of exoplanets by measuring the tiny dip in a star's light as a planet passes in front of it (the transit method). These discoveries have expanded our understanding of planetary systems and have even led to the identification of Earth-like exoplanets in the habitable zone of their stars. The study of exoplanets is important for cosmology because it can help answer one of the most profound questions in science: Are we alone in the universe?

The James Webb Space Telescope (JWST), set to launch in the near future, will take space-based cosmology to the next level. Designed to observe the universe in infrared wavelengths, JWST will be able to peer deeper into the universe than ever before, observing the first galaxies that formed after the Big Bang and studying the atmospheres of exoplanets in detail. With its powerful instruments, JWST will provide new insights into the formation of galaxies, the nature of dark matter and dark energy, and the possibility of life elsewhere in the universe.

In Insightful Reflection, space telescopes have revolutionized our understanding of the cosmos. They have provided us with clear, unobstructed views of distant galaxies, allowed us to map the cosmic microwave background radiation, and offered critical evidence for the existence of dark matter and dark energy. Space telescopes have also helped us probe the fundamental nature of space-time and have contributed to the discovery of thousands of exoplanets. As we continue to push the boundaries of space exploration, these instruments will remain at the forefront of cosmology, unlocking the secrets of the universe and paving the way for future discoveries that will shape our understanding of the cosmos for generations to come.

Chapter 9: The Challenges of Operating Space Telescopes

Operating space telescopes presents a set of unique challenges that differ significantly from those encountered in ground-based observatories. While space telescopes have revolutionized the field of astronomy, providing a clearer and more accurate view of the universe, their operation involves overcoming a range of complex technical, logistical, and environmental issues. From launching these massive instruments into orbit to maintaining them for long-term operations, every aspect of space telescope management demands careful planning, sophisticated engineering, and constant innovation. This chapter will explore the various challenges associated with the operation of space telescopes, shedding light on the intricate processes that ensure these instruments continue to function at the cutting edge of astronomical research.

The first major challenge in operating space telescopes begins long before the telescope even enters space. The design and construction of these observatories require a level of precision that is almost beyond imagination. Telescopes like the Hubble Space Telescope and the James Webb Space Telescope (JWST) are far more intricate than any optical instrument designed for Earth-based observation. Space telescopes must not only have highly sensitive instruments capable of capturing the faintest signals from distant galaxies, but they must also be built to survive the extreme conditions of space. This includes intense radiation, drastic temperature variations, microgravity, and the vacuum of space.

One of the most daunting aspects of designing space telescopes is ensuring that their optics can remain stable and accurate over long periods of time. Even a minor misalignment or distortion of the mirrors could lead to a significant loss of data quality. For instance, Hubble's iconic optics were initially flawed due to an error in the manufacturing of its primary mirror. This problem was corrected by the installation of a corrective optics package during the servicing mission in 1993, but it was a reminder of the extreme precision required when constructing space telescopes. The mirrors and optical components must be finely tuned to avoid optical aberrations that would compromise the telescope's ability to capture high-resolution images.

Once the telescope is constructed, the next major hurdle is its launch. Space telescopes need to be placed into orbit around Earth, often far from the planet's atmosphere, to ensure they can observe celestial objects with minimal interference. The launch itself is a monumental challenge, as the telescopes must withstand the intense forces and vibrations associated with the rocket's ascent. Every space telescope is typically packed into a rocket's payload bay, and the process of releasing it into space involves the delicate deployment of solar panels, antennas, and mirrors. The precise execution of this deployment is crucial; if any components are damaged or fail to deploy correctly, the entire mission could be at risk.

One of the most difficult aspects of launching space telescopes is ensuring that the observatory can maintain its orientation in space. Without the Earth's gravitational pull, the telescope has to rely on advanced gyroscopic systems and reaction wheels to remain locked on its target. These systems ensure that the telescope can maintain its position in the sky with an unprecedented level of precision, often to within a fraction of a degree. Any failure of these orientation control systems can result in the telescope losing its ability to point accurately at specific objects or regions of the sky, severely limiting its ability to gather scientific data.

Once the telescope is in orbit, another set of challenges arises—maintaining the instrument over extended periods of time. Unlike ground-based telescopes, which can be easily serviced or repaired, space telescopes are situated in locations where physical access is impossible. This makes maintaining and repairing them an especially difficult task. While some space telescopes, like the Hubble Space Telescope, have been serviced by astronauts during space shuttle missions, this is not always feasible. For example, Hubble's final servicing mission took place in 2009, and future servicing missions were deemed too risky. As a result, future space telescopes, including the JWST, are designed to be as maintenance-free as possible. However, this approach raises significant concerns regarding the longevity of the instruments, as unexpected malfunctions or failures of crucial systems could lead to a telescope's premature retirement.

Temperature control is another major challenge for space telescopes. Space is an incredibly hostile environment, where temperatures can fluctuate dramatically depending on whether the telescope is in the sun or the shadow of the Earth or the moon. These temperature variations can cause significant expansion and contraction of materials, potentially leading to structural damage or misalignment of the telescope's instruments. To address this, space telescopes must be equipped with sophisticated thermal systems to regulate their temperature. For example, the JWST is equipped with a multi-layered sunshield the size of a tennis court, which protects its instruments from the Sun's heat. This sunshield is crucial for ensuring that the telescope remains at the proper temperature to operate its infrared instruments, which are highly sensitive to changes in temperature.

In addition to the physical challenges of operating a space telescope, the communications systems that allow ground-based scientists to control the telescope and receive data from it must also operate flawlessly. Space telescopes are typically hundreds of thousands, if not millions, of kilometers away from Earth, and their communications with ground stations are subject to the limitations of space-based transmission. Signals from these telescopes must travel vast distances and pass through the Earth's atmosphere, which can cause delays or distortions. Moreover, space telescopes are often placed in orbits far from Earth's direct line of sight, making communication windows limited and sporadic. For example, the Hubble Space Telescope operates in low Earth orbit and can only communicate with ground stations during certain passes, while the JWST will operate at the L2 Lagrange point, where communication with Earth will be less frequent. This means that careful planning and scheduling are required to ensure that data is transmitted efficiently and in real time.

Another significant challenge in the operation of space telescopes is the issue of radiation and its potential effects on the telescope's instruments. Space is filled with various forms of radiation, including cosmic rays and high-energy particles from the Sun. These particles can interfere with the delicate electronics on board the telescope, potentially causing malfunctions or damage. To mitigate this, space telescopes are equipped with shielding to protect sensitive instruments from radiation. However, this shielding is not perfect, and the constant bombardment of radiation over time can still degrade the performance of certain components. Scientists must carefully monitor and assess the impact of radiation on the telescope's instruments, ensuring that they continue to function optimally.

The need for long-term stability and precision in space telescopes also presents challenges when it comes to managing the data they collect. Space telescopes generate enormous amounts of data on a daily basis, as they observe distant galaxies, stars, and other celestial phenomena. This data must be transmitted to Earth, processed, and analyzed in real time, often by teams of scientists and engineers around the world. The sheer volume of data can be overwhelming, and it requires a sophisticated infrastructure of ground stations, data centers, and scientific teams to manage. Data storage and processing capabilities must be constantly upgraded to keep up with the increasing demands of modern space telescopes.

One of the most significant challenges that will be faced in the future is the continued development of space telescopes capable of observing the universe in wavelengths beyond visible light. Instruments designed to study infrared, ultraviolet, and X-ray emissions require specialized technologies and materials that must be developed and tested to the highest standards. As space telescopes venture into increasingly complex scientific territory, the challenges of construction, launch, and maintenance will only grow more demanding. To support these future missions, international collaboration, technological advancements, and careful planning will be essential.

In Insightful Reflection, operating space telescopes involves a complex array of challenges, from the design and launch of these massive instruments to their ongoing maintenance and data management. The extreme conditions of space require precision engineering, while the lack of physical access means that any failures must be dealt with remotely. Despite these hurdles, space telescopes have proven to be invaluable tools for exploring the universe, and their contributions to our understanding of the cosmos are unparalleled. As technology continues to evolve and our understanding of space grows, the operation of space telescopes will remain one of the most exciting and challenging frontiers in modern science.

Chapter 10: Future Space Telescopes: The Next Generation of Observatories

As our understanding of the universe advances, so too does our ambition to explore the cosmos with greater depth and precision. Space telescopes have revolutionized the way we study the universe, pushing the boundaries of astronomical discovery by providing clearer, more detailed observations of distant objects beyond the reach of ground-based instruments. However, the journey of space-based astronomy is far from over. The next generation of space telescopes promises to be even more powerful and sophisticated, ushering in a new era of cosmic exploration. This chapter will explore the upcoming space telescopes, their groundbreaking technologies, and the future challenges and possibilities they bring to the world of space-based astronomy.

One of the key advancements we can expect from the next generation of space telescopes is their ability to explore the universe across a broader range of wavelengths. The first telescopes, like the Hubble Space Telescope, were designed to observe in the visible and ultraviolet spectra, giving scientists a new perspective on cosmic objects. However, as our understanding of the cosmos has evolved, so too has our need to observe in a broader range of electromagnetic radiation. The next wave of space telescopes is designed to operate across infrared, X-ray, and other wavelengths that will allow astronomers to investigate previously hidden aspects of the universe.

The James Webb Space Telescope (JWST) is the most prominent and eagerly anticipated of these new instruments. Set to launch in late 2021 (or early 2022, depending on the latest updates), the JWST is an infrared observatory that will be positioned at the second Lagrange point (L2), located about 1.5 million kilometers from Earth. This positioning will place it in a stable location with minimal interference from Earth's radiation and thermal emissions, allowing it to capture some of the faintest and most distant objects in the universe. Unlike Hubble, which observes mainly in visible and ultraviolet light, JWST will specialize in infrared astronomy, making it ideal for studying distant galaxies, star-forming regions, exoplanets, and the first light in the universe after the Big Bang.

The design of JWST represents the culmination of decades of technological advances. It features a massive primary mirror, which is 6.5 meters in diameter—more than two and a half times the size of Hubble's. The mirror is composed of 18 hexagonal segments made of a special lightweight material known as beryllium, which allows the telescope to maintain its strength while being much lighter than it would be otherwise. JWST's mirrors are coated with a layer of gold to optimize their ability to reflect infrared light. Additionally, the telescope will be protected by a sunshield the size of a tennis court, which will block heat from the Sun and help maintain the telescope's operating temperature at around 50 K (-223 °C), a critical factor for infrared observations.

With JWST's advanced capabilities, scientists will be able to peer into the early universe, observing the formation of galaxies just a few hundred million years after the Big Bang. The telescope will also study exoplanet atmospheres in unprecedented detail, searching for signs of habitability or even life in distant worlds. In essence, JWST will offer an unprecedented look at the universe's history, allowing astronomers to study the birth and evolution of galaxies, stars, and planetary systems.

While JWST is poised to be a monumental leap forward, it is only one part of the broader trend of next-generation space telescopes. Another promising mission currently under development is the European Space Agency's (ESA) *ATHENA* (Advanced Telescope for High Energy Astrophysics). Scheduled for launch in the early 2030s, ATHENA is designed to study high-energy phenomena such as black holes, neutron stars, and supernovae. Unlike previous space telescopes, ATHENA will operate primarily in the X-ray part of the electromagnetic spectrum, which is crucial for studying the most energetic events and objects in the universe.

ATHENA's primary instrument is the Large X-ray Telescope (LXT), which features two nested, multilayer-coated mirrors that will allow for the efficient collection of X-rays. These X-rays are crucial for understanding phenomena such as the behavior of matter in extreme gravitational fields and the interaction of cosmic particles in the vicinity of black holes. ATHENA will be able to observe the hot gas surrounding supermassive black holes and trace the activity of these enigmatic objects, offering new insights into their role in galaxy formation. The mission will also provide data on the high-energy environments of galaxy clusters, allowing scientists to study the distribution of dark matter and its influence on cosmic structures.

Another exciting project on the horizon is the *LUVOIR* (Large Ultraviolet/Optical/Infrared Surveyor), a concept mission being studied by NASA as part of its Astrophysics Probe program. LUVOIR is envisioned to be a multi-wavelength observatory, capable of observing across ultraviolet, visible, and infrared wavelengths. It is designed to be significantly more powerful than both

Hubble and JWST, with a primary mirror ranging from 8 meters to 15 meters in diameter, depending on the specific configuration chosen. LUVOIR will be capable of resolving the atmospheres of exoplanets with an unprecedented level of detail, allowing for the study of potential biosignatures on Earth-like worlds.

The LUVOIR mission is still in the conceptual phase, with the final decision on its development yet to be made. However, if approved, it could be one of the most ambitious and versatile space telescopes ever built. LUVOIR's ability to study exoplanet atmospheres in high resolution would make it an essential tool in the search for habitable worlds, offering new insights into the chemical makeup and potential for life on planets orbiting distant stars. Additionally, LUVOIR's capabilities in the ultraviolet and optical bands would allow for a detailed study of star formation, galaxy evolution, and the interstellar medium.

One of the most exciting aspects of these future space telescopes is their potential to collaborate with other upcoming missions and observatories. For instance, the *Extremely Large Telescope* (ELT) currently being constructed by the European Southern Observatory (ESO) in Chile will be the world's largest ground-based optical telescope when completed. The ELT will complement space-based observatories like JWST by providing a ground-based counterpart that can observe the same objects at higher resolution. The synergy between space telescopes and ground-based observatories will allow for a more comprehensive understanding of cosmic phenomena, as astronomers will be able to compare the observations of the same objects across different wavelengths and from different vantage points.

Another project that will add value to the future landscape of space telescopes is the *Nancy Grace Roman Space Telescope*, also known as Roman. Set to launch in the mid-2020s, Roman will be another infrared space telescope that will build on the legacy of Hubble and JWST. While not as large or as sensitive as JWST, Roman will have a field of view 100 times greater than Hubble's, making it ideal for wide-field surveys of the universe.

Roman's primary science goals include studying dark energy, exoplanets, and the formation of galaxies. The mission will also explore the possibility of detecting faint gravitational lensing signals that could reveal new insights into the nature of dark matter and dark energy.

As space telescopes evolve, they will also increasingly rely on cutting-edge technologies to improve their capabilities. One such advancement is the development of **adaptive optics** for space-based telescopes. Adaptive optics has been a game-changer for ground-based telescopes, correcting distortions in the atmosphere to provide sharper images. While space telescopes are free from atmospheric distortion, they still face other challenges such as thermal effects and mechanical flexing. Researchers are exploring ways to implement adaptive optics for space telescopes to compensate for these issues and enhance image quality.

Another important technological frontier for future space telescopes is the development of **starshade technology**. A starshade is a large, flower-shaped structure designed to block the light of a star while allowing the faint light of an exoplanet to pass through. This technology could be a game-changer in the study of exoplanets, allowing astronomers to directly observe the atmospheres of distant worlds and search for biosignatures. Starshades, when used in conjunction with telescopes like LUVOIR or the Roman Space Telescope, could enable the study of Earth-like planets in the habitable zone of nearby stars.

As the next generation of space telescopes begins to take shape, it is clear that we are on the cusp of a new era in astronomy. These instruments will offer unprecedented views of the universe, allowing scientists to investigate the cosmos in greater detail than ever before. From the study of dark energy and the origins of the universe to the search for life beyond Earth, the possibilities are limitless. However, these ambitious missions will also require continued innovation, international collaboration, and significant investment. As we continue to push the boundaries of our technological capabilities, the future of space telescopes promises to reveal some of the most profound discoveries in the history of science.

In summary, the next generation of space telescopes will represent a quantum leap in our understanding of the universe. With advanced capabilities in multiple wavelengths, groundbreaking technology, and ambitious scientific goals, these telescopes will provide invaluable insights into the cosmos. From studying the earliest galaxies to searching for life on distant planets, the future of space-based astronomy is bright, and the next few decades will undoubtedly usher in a golden age of cosmic discovery.

Chapter 11: Space Telescopes and Citizen Science

The relationship between space telescopes and citizen science has become an increasingly important facet of modern astronomy. Traditionally, space missions were reserved for trained professionals in the scientific community, with data collected from complex instruments analyzed by experts. However, over the past few decades, the rise of citizen science—where non-professionals contribute to scientific research—has begun to make a significant impact in astronomy. The combination of space telescopes' advanced capabilities and the engagement of citizen scientists has opened up new possibilities in astronomical research, enabling more people to participate in the discovery and exploration of the universe.

Citizen science is not a new concept, but the digital age has transformed it into a powerful tool for scientific advancement. Through online platforms and projects, amateurs can contribute to research in ways that were once unimaginable. In the context of space telescopes, this means allowing people from all walks of life to analyze images, identify celestial objects, and even participate in discovering new phenomena that would otherwise go unnoticed. With space telescopes like the Hubble Space Telescope, the James Webb Space Telescope, and others, data is continuously collected and made publicly available, providing ample opportunities for citizen scientists to contribute.

Space telescopes generate vast amounts of data, much of it in the form of images captured from deep space. For example, Hubble's observations of distant galaxies, nebulae, and black holes result in enormous image files, which require hours of analysis to interpret. The complexity of these images—sometimes involving millions of stars, galaxies, and intricate structures—makes it difficult for astronomers to process them all on their own. This is where citizen science comes into play. By leveraging the power of crowdsourcing, space telescopes can tap into the skills and enthusiasm of the global public, leading to the identification of interesting objects, the discovery of rare phenomena, and even new ways of analyzing data.

One of the most prominent examples of space telescope citizen science in action is the *Hubble's Hidden Treasures* project. In this initiative, amateur astronomers and astronomy enthusiasts were given the opportunity to sift through the extensive archive of Hubble images and identify hidden gems—such as rare nebulae, galaxies, and star clusters—that had not been fully analyzed by professional astronomers. Participants were invited to download and explore these images, using their keen eyes to spot objects that might have been overlooked in the vast dataset. Some of these amateur discoveries were so impressive that they even resulted in follow-up studies by professional astronomers.

The success of this project was largely due to the ability of citizen scientists to bring fresh perspectives and creativity to the task. What professional astronomers might overlook in their systematic analysis could be noticed by an amateur who has a unique approach or perspective on the data. In many ways, citizen science brings an element of spontaneity and diversity to the scientific process, increasing the chances of important discoveries. Hubble's Hidden Treasures is just one of many citizen science projects facilitated by space telescopes.

Another exciting example of citizen science involving space telescopes is *Galaxy Zoo*, a project that began as part of the Sloan Digital Sky Survey (SDSS) but has since expanded to include data from other space telescopes like Hubble. The goal of Galaxy Zoo is to engage the public in the classification of galaxies. The sheer number of galaxies in the universe makes it nearly impossible for professional astronomers to classify them all, but by involving volunteers, the project has achieved a level of coverage that would have otherwise been unattainable. Participants are asked to categorize galaxies based on their shape (spiral, elliptical, irregular) and other features. In doing so, they help astronomers classify galaxies in ways that inform our understanding of their evolution and structure.

The success of Galaxy Zoo has been profound. Since its inception, the project has contributed to numerous scientific papers and discoveries, including the identification of unusual galaxies and the development of new ways to understand galaxy formation. What's particularly noteworthy is the sheer scale of participation. Hundreds of thousands of people from all over the world have contributed to this project, demonstrating the power of global collaboration in scientific discovery.

The popularity of Galaxy Zoo has also led to the development of other projects within the same vein, each utilizing space telescope data to address different questions in astronomy. For example, *Supernova Zoo* engages citizen scientists in the identification of supernovae, while *Star Counts* allows the public to classify stars within our galaxy. These projects, which often use data from space telescopes like Hubble, contribute to a better understanding of the universe and provide an avenue for individuals to make meaningful contributions to the scientific community.

The role of space telescopes in citizen science is not limited to image analysis. With advancements in machine learning and artificial intelligence, space telescopes are increasingly generating vast amounts of data that require sophisticated processing to identify patterns and objects of interest. In many cases, citizen scientists can play a role in training algorithms to recognize specific phenomena or objects. For example, the *Planet Hunters* project encourages volunteers to analyze light curves from space telescopes like Kepler, which measure the brightness of stars. By identifying periodic dips in brightness, citizen scientists can detect exoplanets transiting across their stars—an important technique in the search for new worlds.

With the upcoming launch of the James Webb Space Telescope (JWST), the potential for citizen science to contribute to astronomical research will grow exponentially. JWST's unprecedented ability to observe distant galaxies and exoplanets will provide vast amounts of data that will need to be analyzed in real-time. As part of the public outreach for JWST, NASA has already launched citizen science programs designed to engage the public in the analysis of the telescope's observations. For example, the *JWST Citizens Science* project will invite volunteers to help identify and classify new galaxies, stars, and other cosmic objects detected by the telescope. This will allow people from all over the world to actively participate in one of the most ambitious space missions in history.

Beyond identifying objects and classifying data, citizen scientists have also contributed to the development of new tools and methods used in space astronomy. Open-source software and platforms developed by the community have been instrumental in enabling data processing and analysis. These tools are often shared across scientific communities, allowing researchers from different disciplines to collaborate and build on one another's work. In some cases, citizen scientists themselves have gone on to develop software or algorithms that have become integral parts of the professional astronomy workflow.

The integration of citizen science with space telescopes also presents unique educational opportunities. For many participants, contributing to space telescope projects provides a hands-on, immersive experience in science and astronomy. This involvement not only enhances public understanding of the cosmos but also inspires the next generation of astronomers and scientists. By making space research more accessible, space telescopes allow individuals to engage with scientific inquiry in ways that were once reserved for specialists. This democratization of knowledge fosters a deeper appreciation of science and encourages curiosity-driven exploration.

However, the intersection of space telescopes and citizen science does not come without challenges. One of the primary issues is the need for effective data management and quality control. While citizen scientists can make important contributions, the sheer volume of data and the complexity of some tasks can lead to errors or inconsistencies. To mitigate this, many citizen science projects use multiple participants to review the same data, ensuring that findings are accurate and reliable.

Furthermore, advanced algorithms and machine learning techniques are being developed to assist citizen scientists in their analysis and to help filter out false positives.

Another challenge is the need to maintain the enthusiasm and participation of volunteers over time. Citizen science projects often rely on public engagement and sustained interest, and the complexity or length of a project can deter continued involvement. To address this, many space telescope citizen science projects offer incentives such as digital badges, certificates, or recognition on their websites. Additionally, projects that emphasize community involvement, collaboration, and the excitement of discovery tend to attract dedicated participants who remain engaged over the long term.

Despite these challenges, the potential benefits of citizen science in space astronomy are enormous. By harnessing the collective power of a global community, space telescopes can accelerate the pace of discovery, increase the amount of data analyzed, and provide new insights into the universe. Citizen scientists, whether amateurs with a passion for space or students looking to deepen their knowledge, are making invaluable contributions to the study of the cosmos. As space missions like JWST and the upcoming LUVOIR telescope come online, the role of citizen science will only grow, cementing its place as an integral part of modern astronomy.

In Insightful Reflection, the relationship between space telescopes and citizen science is one of mutual benefit. Space telescopes provide the tools and data needed for discovery, while citizen scientists contribute their enthusiasm, creativity, and observations. Together, they have the potential to unlock new mysteries of the universe and reshape our understanding of the cosmos. As more people around the world engage with these exciting projects, the future of space exploration will become even more inclusive, collaborative, and accessible to all. Through citizen science, we are all becoming explorers of the universe.

Chapter 12: The Impact of Space Telescopes on Humanity

The role of space telescopes in our understanding of the cosmos goes far beyond merely providing us with awe-inspiring images of distant galaxies, nebulae, and stars. They have fundamentally altered our view of the universe, reshaping our place within it and influencing many aspects of human society. Through their advanced technology and the scientific discoveries they enable, space telescopes have made profound impacts not only on the field of astronomy but also on human culture, our philosophical outlook, and the way we approach scientific inquiry. Their influence extends to education, technological innovation, international collaboration, and even our conception of the very nature of existence. This chapter explores how space telescopes have shaped and will continue to shape humanity in ways we may not fully yet understand.

The most immediate impact of space telescopes is the way they have expanded our understanding of the universe.

Space-based observatories like the Hubble Space Telescope, Chandra X-ray Observatory, and the James Webb Space Telescope (JWST) have allowed humanity to peer deeper into the cosmos than ever before. They have provided detailed, close-up images of distant planets, stars, and galaxies, uncovering phenomena that were previously hidden from view. For instance, the Hubble Deep Field image, which was captured by the Hubble Space Telescope in 1995, revealed thousands of galaxies in a small patch of sky, offering an unprecedented glimpse into the distant universe. This image, along with many others, forced humanity to rethink our place in the universe. We are no longer just observers on a pale blue dot in a vast cosmos; we are participants in a much grander story of cosmic evolution.

One of the most significant revelations from space telescopes is the discovery of exoplanets—planets that orbit stars outside our solar system. The Kepler Space Telescope, for example, revolutionized our understanding of the frequency and diversity of exoplanets, leading to the identification of thousands of these distant worlds.

Space telescopes have enabled scientists to observe exoplanets in ways that were previously impossible, including studying their atmospheres and conditions. This has sparked one of the most exciting areas of research in modern astronomy: the search for extraterrestrial life. As scientists identify more Earth-like planets in the habitable zones of their stars, the possibility of life beyond Earth becomes more plausible. This question—whether we are alone in the universe—has profound philosophical and existential implications for humanity, challenging our notions of life, intelligence, and the uniqueness of our existence.

Beyond their scientific discoveries, space telescopes have also had a profound cultural impact. Images captured by observatories like Hubble and JWST have become iconic in their own right, captivating the public's imagination and inspiring countless individuals to pursue careers in science and technology. The ethereal beauty of these images—swirling nebulae, majestic galaxies, and explosive supernovae—have introduced millions of people to the grandeur of the universe. These images have also served as reminders of the fragility of our planet and the need to preserve our environment. The Earthrise photograph taken by the Apollo 8 astronauts in 1968 had a similar effect, awakening a collective environmental consciousness. The images from space telescopes continue this tradition, reminding us of the vastness of the universe and the preciousness of our home.

Space telescopes have also had a significant influence on the way we view our own planet. In many ways, they have underscored the uniqueness and vulnerability of Earth. As we gaze upon the swirling clouds of Jupiter or the barren surfaces of distant exoplanets, we are reminded of how special and rare our own world is. The view from space is humbling, and it fosters a sense of unity among humanity. It has become increasingly clear that we share a common destiny on this fragile planet, and the insights provided by space telescopes only reinforce the need for global cooperation in addressing the challenges facing humanity, from climate change to the peaceful use of space.

The impact of space telescopes extends into the realm of technological innovation. The technology developed for space observatories has led to numerous breakthroughs in other fields, from medicine to environmental monitoring. For instance, the Hubble Space Telescope has contributed to the development of advanced imaging technologies, such as adaptive optics, which are now used in a variety of applications, including eye surgery. The data processing techniques developed to handle the enormous volumes of data collected by space telescopes have also influenced fields as diverse as communications, computer science, and remote sensing. The process of developing space-based telescopes requires solving complex engineering challenges, leading to innovations that have far-reaching consequences for many industries. In essence, space telescopes act as incubators for technological progress, advancing our understanding of the universe while simultaneously driving technological development that benefits society.

Another profound impact of space telescopes is their role in fostering international collaboration. The development, launch, and operation of space telescopes are not the work of a single nation or organization. They are typically the result of joint efforts between governments, research institutions, and private companies from around the world. The International Space Station, for example, has been a model of international cooperation, with space telescopes like Hubble and JWST benefiting from similar collaboration. The Hubble Space Telescope was launched and is operated through the combined efforts of NASA, the European Space Agency (ESA), and the Canadian Space Agency (CSA). This spirit of cooperation is not just a political necessity but also a philosophical one, as space telescopes remind us that we are all citizens of the same planet. In an era of geopolitical tensions, the shared pursuit of space exploration serves as a beacon of hope for peaceful collaboration.

Space telescopes also have a profound effect on education. By making their data publicly available, space telescopes have created countless opportunities for educational outreach, providing students and educators with rich, real-world data for scientific analysis. Programs like NASA's *AstroGraphics*, which allow students to work with actual Hubble Space Telescope images, and *NASA's Eyes on the Solar System*, which enables users to explore the universe in three dimensions, offer hands-on opportunities for students to engage with space science in an interactive and dynamic way. These educational initiatives inspire the next generation of scientists, engineers, and space enthusiasts, nurturing a curiosity-driven culture of inquiry that transcends national boundaries. The data and images produced by space telescopes provide a vast resource for educators, enabling them to bring complex scientific concepts to life in the classroom.

The impact of space telescopes also stretches into the realm of philosophy and human consciousness. As we study distant galaxies and the oldest light in the universe, we are confronted with questions that go beyond the realm of science. What is the nature of time and space? How did the universe begin, and what is its ultimate fate? The discoveries made by space telescopes have given us new tools to explore these existential questions, but they have also raised more questions than they have answered. The images of the farthest reaches of the universe challenge us to think about our place within the cosmos, forcing us to confront the idea that we are just a tiny part of an enormous, ever-expanding universe. In this sense, space telescopes invite us to consider not only the nature of the universe but also the nature of humanity itself.

Finally, the study of space through telescopes has created a deep connection between humanity and the universe, one that transcends cultural, religious, and political differences. Space exploration and the pursuit of knowledge about the cosmos are universal endeavors that belong to all of humanity. When we look up at the stars, we are reminded of our shared heritage and the infinite potential of human ingenuity. The quest to understand the universe is, at its core, a quest for meaning, and space telescopes have become one of the most powerful instruments in that search. They provide us with the tools to explore the deepest mysteries of existence and challenge us to think beyond our immediate surroundings.

In Insightful Reflection, space telescopes have had an immeasurable impact on humanity. They have expanded our scientific understanding, influenced our culture, and provided a new perspective on the universe and our place in it. Through their discoveries, they have sparked new philosophical questions and fostered international collaboration. They have spurred technological innovation and enriched educational experiences. Most importantly, they have reminded us of the vastness of the cosmos and our fragile, interconnected existence on Earth. The legacy of space telescopes is not just found in the data they produce or the discoveries they enable, but in the way they have transformed the human spirit, inspiring wonder, curiosity, and the pursuit of knowledge. As we continue to explore the universe, space telescopes will undoubtedly remain at the forefront of our journey into the unknown.

Printed in Great Britain
by Amazon